# AUGMENTED INK

How Writers, Artists, and Creators Can Use AI
to Elevate, Not Erase, Their Voice

## MICHELLE JESTER

ISBN: 978-1-964026-11-4 (paperback)
ISBN: 978-1-964026-12-1 (ebook)

AUGMENTED INK: How Writers, Artists, and Creators Can Use AI to Elevate, Not Erase, Their Voice

MH
MICHTER HOUSE
PUBLISHING
an imprint of
Rope Swing Publishing

www.ropeswingpublishing.com

For all writers, artists, and creators!

# INTRODUCTION

*AI is not the enemy of creativity...*
*but it is not the creator, either.*

We are in the age of accessible artificial intelligence, whether any of us like it or not. As an author and publisher, I've witnessed the rise of it in the creative world from both sides of the desk. I've seen authors, music producers, and content creators use it brilliantly, treating it as a tool to support brainstorming, enhance clarity, and streamline workflow. I've also seen it used very poorly, allowing machines to fully take the place of voice, vision, and effort. The truth is, AI is neither inherently harmful nor inherently helpful. It is simply a tool. Much like a sword, it can be extremely useful, but its value depends on how and why we use it.

My dad always loved technology and was fascinated by the growth of it. As soon as something new was released, he bought it. Betamax, VHS, MCA DiscoVision, cassette recorder, Sony Walkman, and the microwave oven come to mind first. When home computers first became available, we had one. The Commodore 64.

Technology was a major part of my childhood.

I remember vividly taking a class as a family on how to use the microwave. I remember taking trips to one of the few stores that carried large LaserDisc movies and purchasing several. My sister and I wore out Grease from playing it over and over and over. Some of my fondest memories is watching my dad listen to his classical music on his Walkman while he worked in the garage or yard.

My sister and I ate up the technology too, especially the computer. My sister went on to major in computers in college, one of the few that offered such a thing back then. We were never afraid of technology; instead embraced it and enjoyed it. My dad instilled in us a love and fascination for growth, but also an understanding of where that growth might take us. Yes, even back then, my dad said, "You watch, one day, these things will be doing all the work for us."

Still, he was unafraid to enjoy technology and be one of the first to buy in.

When the movie *Wargames* was released, he said, "Girls, this will likely happen." Back then, watching *Wargames*, it was impossible to think a computer could talk aloud, even in the awkward computer-like tones as in the movie. Now Alexa and Siri do so with a simple prompt or question. We can also generate realistic-sounding voices simply by plugging in a few words. Each of these technologies have brought fear and worry.

With every forward movement in technology, there are always going to be concerns.

I understand the appreciation, as well as the apprehension.

The Betamax brought not only fears

that piracy and unauthorized copying would run the entertainment industry out of business but also worries over the fact that people could record and fast forward through commercials (which we absolutely did) would lead to advertising revenue losses. Advertisers were anxious that their ads would be devalued or useless. Lastly, the entertainment industry had major concerns over the extreme impact Betamax would have on theatrical releases.

The "Betamax Case" attempted to block the technology. The landmark lawsuit brought by Universal Studios and Walt Disney against Sony in 1976 aimed to halt the sale of Betamax machines. They argued that Betamax technology facilitated copyright infringement.

A district court ruled in Sony's favor in 1979. The case's ultimate resolution in 1984 came with the Supreme Court's decision in favor of Sony as well.

The case highlighted the industry's significant concerns about the potential consequences of new technology on the entertainment companies overall. These concerns over the years, through every jump in technology, from VHS, Compact

Disc, DVD's and streaming services, are nothing new.

In every instance, the industry overcame these obstacles by adapting its business models. Creating new revenue streams, regulating use, and ultimately creating demand, which in turn provided more jobs.

Technology advancements, including CGI and virtual reality, have dramatically altered filmmaking techniques and created new avenues for entertainment experiences as well as opened up new career paths. There has been an increased production of content fueled by the internet and digital platforms, and the entertainment industry is creating and distributing more content than ever before.

Stardom is no longer limited to movie and TV stars and now includes influencers and online personalities. Many people who never had a chance to get their works out to the public now have a space.

Jumps in the tech industry have also created a larger space for many new entertainment avenues. More movies, music, actors, artists, editors, and every other career associated with the industry.

It is evident that it took away a bit of control from the few that had been prevalent for years, which is never comfortable, but again they adapted and more artists, actors, and authors that wanted a seat at the table, now have one.

Many of the fears, all the way through advances in technology, have been overcome in the same manner: adapting responsibly.

In this book, I will often be addressing the specific area of writing, although I will be including other areas of the entertainment industry as well. I do not intend to dissuade creators from using AI. On the contrary, I believe AI has real potential to support many areas of the entertainment process. It can speed up production and relieve some of the mental load that comes with juggling what are called the "professional demands." Used wisely, AI can help writers, authors, and creators stay productive and even become more confident in their craft as they focus on the actual creative process.

It's clear from the facts that AI cannot provide depth by itself. That still takes a person behind technology, as always. It cannot speak from personal experience or feel heartbreak, wonder, regret, or hope. It cannot draw meaning from memory or emotion from stillness. What it produces may *look* like creating, but without the heart and intention of a human behind it, it rarely resonates. It can mimic, but it cannot live.

This book is based on a booklet I wrote for the publishing company in the early stages of AI use. The more we started recognizing manuscript submissions with less heart and more flat content, we knew we had to address it. This is for the writer who is curious about using AI but doesn't want to lose their voice in the process. The one who is fascinated but also cautious. It's for the teacher guiding students in this rapidly growing landscape. It's for the small business owner, the coach, the content creator, music producer, and the storyteller who wants to work smarter, and also stay rooted in what makes their message matter.

My hope is that this book will help you understand how to use AI to support you in your creative process.

The world still needs you, and AI, if invited in wisely, can be part of the process without ever becoming the meat of it.

# CHAPTER 1

## *The Rise of the Machine*

Not too long ago, the notion of a machine helping us write a book, or even a grocery list, felt like a science fiction subplot. Writers have long been cast as lone creators, curled over notebooks or keyboards, bleeding onto the page. The idea that something artificial, something coded, stepping into that sacred space felt like a threat. A violation. But the world has shifted, as it has many times in the past. With leaps in technology. Alexa and

Siri can both create a shopping list for you as well as email it to you and others. As artificial intelligence emerged not as a future concept, but a present reality, one capable of reshaping the very act of creation, I felt led to write this book.

The naysayers who strongly oppose AI technology have likely already been using and consuming it for years. Every time we accept autocorrect on a text message, rely on everyday programs to smooth a sentence, let predictive search finish our thought, or suggest a song, we've invited AI into our world. We use AI face apps, game cheats, and a host of other computer-generated assistance. What many people don't realize is that AI has been around for many years—decades. "Artificial intelligence" was coined at the 1956 Dartmouth conference, marking the formal beginning of AI as an academic field in the mid-1950s. However, AI's foundation stretches back centuries, built upon ancient ideas and later mathematical and computational progress. While it wasn't as progressive as it is now, it's still been a part of our lives and it's not going away.

What's new, and what's alarming to

some, is how much further that assistance can now go. In a matter of seconds, AI can generate essays, stories, outlines, and even poetry. It can simulate voices, mimic tone, and regurgitate the patterns of great authors past. It can create images, videos, and music compositions. For some, this feels like a liberation. For others, a slow, excruciating creative death.

We've seen an increasing, and to be honest, an alarming uptick in manuscripts being submitted using the majority of AI content. If you are asking how we can tell? There are signs... many signs. Still, AI isn't going away. So, the real question isn't whether writers should use it. The question is how it *should* be used.

There's a seductive ease in letting a machine do all the work. A blank page, while it can be exciting for some, it can be terrifying for others. That doesn't mean AI has no place in the process. On the contrary, as with cassette or VHS recorders, when used wisely and ethically, AI can be a remarkable partner, making our lives more enjoyable. The danger lies not in its presence, but in the temptation to misuse it.

At its best, AI serves as an extension

of the creator's mind. For writers, it can help explore alternate plot twists, offer fresh phrasing, or help tighten structure. It can assist with tone, summarization, brainstorming, and outlining. It can ask questions you haven't considered. What it cannot do, at least not yet, and arguably never with human depth, is bring *emotional truth* to the page. It cannot draw from your lived experience. It doesn't know the pain behind your prose or the joy behind your humor. It cannot see the room you wrote in, or the years you carried the story before you dared to write it.

That matters.

The fear among many creatives is valid. Will publishers replace writers with prompts and algorithms? Will originality vanish beneath a flood of machine-made mediocrity? Will the art of writing, of sweating through a first draft, agonizing over word choice, and cutting what you love, be replaced by something faster, but emptier?

Some of the same questions apply to other areas of the entertainment industry. Will music and film engineers, filmmakers, and illustrators be replaced with AI-generated content?

These are the right questions. But fear, when held without vision, tends to lead to retreat. The better path is adaptation. Understanding. Even partnership.

I'm not writing a love letter to AI, by far. Nor a eulogy for human creativity. I'm just having a conversation, an invitation to explore how we can remain rooted in our own voice while learning to use the tools that are changing the world. Technology has always shaped writing and creative content. The first color camera changed movies. Machine learning (ML), and specialized software changed animation. The printing press changed book and publication distribution. The typewriter changed speed. Word processors changed editing. The internet changed research and reach. Now, AI is changing everything else.

A phrase I used often in conferences, and my own writings, albeit about marriage, is that *you will either grow with your spouse, or they will outgrow you. Because things change, people change, the world changes.* Same principle with technology, it changes, and you will grow with it, or it will outgrow you.

Many creatives get stuck in their work by not having access to the tools they need. A songwriter without the composition or vocals, or a singer without a song or music producer are halted in their creative progress.

For many creators, especially those outside traditional systems, AI is opening doors that have been closed to them for a long time. They can generate variations, summaries, or even marketing copy. Tasks that once drained their time and energy are now readily available. For indie authors, small presses, and self-publishers, AI is leveling the playing field. To them, it's not about cheating, it's about choice.

Still, choice requires clarity. Just because AI can write for you doesn't mean it should. Just because it can edit, AI still can't do what a storyline editor can on an emotional level. Just because it can help with music composition, doesn't mean engineering to polish the work won't be needed. So, we must draw personal lines.

Some may use AI to brainstorm dialogue but insist on crafting the final scene themselves. Others might lean on it to summarize chapters but never touch a single line. AI must be utilized with intentional use. AI should serve the process, not replace it. The writer still must be the writer. The songwriter must still pour into their song. The graphic designer must still edit an image or advertisement. Because all of these creators are still the ones telling the story.

In fact, the best uses of AI often lead to deeper creativity. It can help you break through blocks, identify inconsistencies, and rethink phrasing. It can offer options one might not have considered. Think of it like a brainstorming partner who never judges, never sleeps, and always has something to say. Of course, not everything it says will be good, or even make sense, and especially not always accurate. But that's part of the process. Your job isn't to accept everything AI offers; *your job is to curate, to discern, and to lead.*

Writers, artists, and creators must resist the urge to swing to extremes. AI is neither the villain nor the savior. It is a tool, one that reflects what you ask of it. If you feed it shallow prompts, you'll get shallow content. If you engage with curiosity and critical thinking, it will reflect that back. You have to stay present. You must stay in the creative seat. The moment you hand over your vision to a machine, it stops being art and becomes automation.

Writers are often told to "write what you know." But writing is also about discovery. It's about asking what if, exploring what you don't yet understand. Every single author I've ever met grew personally with every book *they* wrote. AI, when used ethically, can enhance that journey. It can be a flashlight in the dark, not a GPS that tells you where to go, but a lamp that illuminates the path so you can choose your own steps.

Let me say there's something very humbling in working alongside AI. It reminds us that creativity isn't about control, it's about communication. Our story doesn't exist in a vacuum. It's shaped by tools, culture, language, and audience, and has for years. AI is now a common part of that landscape. We don't have to love it, but to ignore it is to fall behind. Worse, it's missing an opportunity, to reclaim your time, sharpen your craft, engage in new ways to get your voice into the world, and have a new kind of conversation between mind and machine.

I wrote this for every writer and creator to remind them to stay human in an increasingly digital world. For the creatives who want to use technology without being used by it. For those who believe that voice still matters, even if algorithms now share the stage. Most of all, it's for anyone who feels torn between curiosity and caution and needs permission to explore both.

This is not the end of entertainment. It's the next evolution. The pen, mouse, keyboard, and microphone are still in your hand. AI might offer a map, but you're still the one choosing the road. So, do it boldly. Remember, while AI can generate text, music, and images, only you can tell a story that breathes.

And like my dad, don't be afraid of technology, be mindful of it.

# CHAPTER 2

## *The Human Touch Still Matters*

It is easy to forget as artificial intelligence becomes more capable and integrated into our world, that behind every great story, compelling argument, or moving piece of art, there is a human hand guiding the work. There is a human voice shaping tone, choosing structure, and delivering meaning. Despite the impressive evolution of generative tools, the emotional truth of artful storytelling remains deeply human.

It is this lived quality that no machine can replicate, not fully, and not authentically.

AI excels at mimicking patterns. It has been trained on vast amounts of text, music, and imagery, learning how they are likely to appear next to one another, recognizing structure, rhythm, and genre. It can craft a poem that looks like a poem or write a plot twist that feels familiar. But while AI can generate content that passes a casual glance, it cannot *know* heartbreak. It cannot remember what it felt like to be ten years old and left out at recess, or to stand beside a parent's hospital bed and feel time slip away. These moments, stitched into memory and feeling, are where real creating finds its power.

Humanity creates not only to communicate, but to process, to question, to grieve, and to celebrate. I write as an outlet. I have since I was ten years old. Poems, songs, short stories, novels, and personal reflections, I write all of it. Whatever hits me, that's what I do. Many people write to understand themselves and each other. To express and sometimes relieve ourselves of burden.

When you read something that

moves you, it is not the cleverness of the sentence alone that draws you in, it is the truth behind the word. The sincerity and vulnerability of a mind opening itself to you. AI can generate imitation, but it cannot feel. While some might argue that feeling is not required for effective writing... readers, editors, and publishers instinctively know the difference between writing that was lived and writing that was simply produced.

This is not to say that all human creations are emotionally profound or that all AI creation lacks soul. Rather, it is to recognize that true connection in stems from experience. A song written from a place of real longing, a memoir shaped by years of reflection and pain, or a novel born of tragedy carries something AI cannot fabricate.

My husband and daughter are both painters. They each have different styles, but in every work they produce, one can feel the emotion behind it. This nuance of emotion, the unspoken tension in one of the strokes or drawings, hint at deeper wounds. These are not just technical choices, they are human ones.

In a world increasingly filled with

machine-generated content, human authenticity *becomes the competitive advantage.* Truly. Not speed or replication. In publishing, readers are growing savvier as more and more AI books hit the market. When we scroll through reels, watchers are quick to point out AI video content. Many people can often sense when something is hollow, when the voice is too mechanical, too generic, too sanitized, or on the opposite end when something is too outlandish, too perfect, or too smooth. The imperfections of human creativity, its rawness and occasional awkwardness, are often what make it compelling.

Ever read a review that says something like, "I have always loved his writing, but this one fell short. Is he using AI now? Because I *felt* none of the conflict."

In writing, AI may be able to generate a motivational blog post or summarize a dense article, but it cannot write from the perspective of a woman rebuilding her life after loss, or a young man grappling with identity in a culture that silences him. It cannot imagine the smell of a childhood kitchen or the ache of a long goodbye in the same way a person can. It does not carry precious memories, and it holds no hope. These are human capacities, and they are the very heart of real creating.

There is also the matter of voice. Every great writer has a voice that is unmistakably theirs. It is not just about vocabulary or sentence structure; it is about rhythm, humor, cadence, and worldview. Sometimes, it's about crutch words or phrases (I have a few) none of which typically bother the reader. Editors are good at taking out overused words while still leaving enough of them to reflect the writer's true voice. It is about how a person sees the world and how they choose to translate that vision into language. AI can simulate voice, but it does so based on patterns it has already seen. It does not invent from the soul; it assembles from statistics. Your voice, by contrast, is shaped by your experiences, your upbringing, your failures, your faith, your fears, your heritage, and your hopes. No one else can write exactly as you do, not even an algorithm trained on billions of words.

Some writers may worry that their voice is not strong enough or unique enough to matter in the age of AI. This is a valid concern, especially in a culture that often rewards speed over depth and volume over quality. However, *it is precisely this saturation that makes human voice more valuable.* In a landscape cluttered with efficient sameness, authentic difference stands out. Readers are not just seeking content... they are seeking connection. They want to hear from someone who has lived, someone who has questioned, someone who is trying to make sense of the world in a way that *feels* real.

The creative process, too, is more than the products it produces. There is something sacred in the struggle, in the drafting and redrafting, in the silent hours spent chasing a sentence or drumbeat that refuses to land. I know this well. In my creative process, so many of those moments taught me patience, discipline, and self-trust. I was often able to work out my feelings and disappointments simply by writing them down. AI offers convenience, but it does not offer personal growth. It cannot challenge you to confront your own biases, to refine your thinking, or to dig deeper into the truth of your own narrative. It cannot ask, "Why does this matter?" or "What are you really trying to say?"

Creating has always been an act of discovery. With writing, often the story leads us somewhere unexpected, forcing us to face parts of ourselves we had not intended to examine. This exploration, this movement through uncertainty, is central to the craft. AI can assist with structure or provide options, but it cannot take that internal journey for you, nor can it provide that same journey for the reader.

During the time I was writing the novel *It Never Occurred to Her*, I would be at my computer weeping as I typed. My husband grew concerned one day, came into the study and asked me to step away from writing for a few days. He was concerned. But I couldn't. I was in the story so much that I could feel every emotion, and I poured each one onto the page. I did end up taking a break for a few days, but only to regroup, refuel for the rest of the story. Whether writing fiction or nonfiction, that's the writing process. It often hurts, can sometimes feel like you are learning something new yourself, and that's good. Yes, it's emotionally draining, but when the work is done, it is a fully formed work of art that you've poured your humanity into. Sometimes taking that break is the actual breakthrough you needed to complete the work.

It is tempting to chase efficiency, especially when deadlines loom and attention spans shrink. But (notice one of my crutch words yet?) we must resist the idea that faster is always better. Writing is not just about productivity. It is about presence. True, emotional cathartic presence. It is about sitting with an idea long enough to understand it, shaping it carefully, and releasing it into the world with integrity. That process is part of the value of the work.

This does not mean that AI has no place in the creating process. Used with discernment, it can be a helpful tool, a brainstorming partner, an initial editor, a prompt generator, an inspiration muse. It can make some aspects of the process easier, freeing up mental space for deeper creative work. However, we must always remain in control. We must be the final voice, the final hand, and the final brain behind the result. Otherwise, we risk becoming curators of content instead of creators of meaning.

So, (another favorite of mine) as AI

continues to evolve, let us evolve with it, not by surrendering, but by standing firm in what makes our work worth creating.

Let us remember that the value of it lies not only in what is done, but in who is doing it.

Finally, let us trust that in a world of automation, authenticity will always find its audience.

# CHAPTER 3

*Writing with a Co-Pilot: Best Practices*

The idea of writing with artificial intelligence was something reserved for the realm of speculative fiction or experimental laboratories. Today, however, it is increasingly common. Writers across genres and industries are using AI to assist with everything from idea generation and content outlining to proofreading and developmental edits. As with any tool, the value of AI in the writing process depends not just on what it can do, but on how intentionally it is used. Learning to work with AI as a co-pilot, rather than letting it take over as a driver, requires thoughtfulness, boundaries, and a deep understanding of one's own goals.

First, it is helpful to acknowledge what AI does well. It excels at pattern recognition, summarization, rapid content generation, and structural suggestions. It can quickly create lists, rewrite paragraphs in a different tone, reframe composition, or create outlines. For writers who face tight deadlines, creative fatigue, or cognitive overload, these capabilities can be a welcome relief. AI can break through blocks and reduce the strain of repetitive tasks, offering a starting point that would otherwise take hours to shape manually. If you use it right, it's like a helpful partner that boosts your work without stealing your originality.

We need to remember that productivity isn't the same as creativity. I had an author who let me know that she'd written four more books since we last spoke the week before. I replied, "You have not *written* four books since last week. You may have four new books, but I can absolutely, without a doubt, state that *you* didn't write them." She paused before admitting that she'd generated them based on summaries she'd previously written. My next question was, "Have you read, edited, added to, or rewritten any of the content? Does the book encompass your entire story and everything you thought about? Because I know you couldn't have done that working a full-time job, in a week. We have no issues with people using technology, but we do require transparency and originality. So, for that reason, we're not interested." She was devastated, because she felt accomplished. However, nearly seven weeks later, when she called again, she said she would've been devastated had she submitted one of those books to us. They might have had her basic truths in them, but they did not have her story. She is currently in the process of reading through and editing the first one, and we look forward to reading what she produces.

This is what I am trying to get across. While AI can help spark ideas, it does not originate them in the same way a human mind does. Yes, it draws from your prompts or summaries, but it cannot inject your voice and truth. It draws from existing material, weaving patterns based on what it has already encountered. Therefore, when working with AI, the creator must remain the source of direction, intention, and authenticity. Asking a tool to generate a blog post on gratitude, for example, will yield common phrases and accessible language. However, unless personal perspective and voice are injected into the result, the piece will likely feel generic, lacking the human quality that distinguishes memorable writing from mere information.

One of the best ways to use AI as a co-pilot is in the early stages of writing, during brainstorming, outlining, and structuring. At this stage, the goal is not perfection but movement. Many writers struggle to begin because the blank page is intimidating. AI can lower that initial barrier. By suggesting offering outlines, or listing potential subtopics, it gives the writer something to respond to and tweak. These responses, even when imperfect, provide momentum. They create friction against which original thought can take shape. The process becomes less about staring into the void and more about shaping raw material into something purposeful.

Writers can also benefit from AI during the revision stage. After drafting a chapter, article, or essay, it is common to feel too close to the work to assess it clearly. AI can offer a second set of eyes, pointing out unclear phrasing, repeated words, or overly complex sentences, before it ever goes off to an editor. Some tools allow you to ask for alternate versions of a paragraph or different transitions between sections. These suggestions are not always superior, but they serve as useful comparisons, helping the writer evaluate options with greater clarity in the re-writing stage. In this way, AI becomes a sounding board, not a dictator of style.

Still, boundaries must be established. When using AI, it is important to be clear about which parts of the work you are comfortable outsourcing and which you consider non-negotiable. For example, some writers use AI to generate titles, taglines, or metadata but draw the line at full sentence generation. Others may use AI to draft a rough introduction, then rewrite it entirely in their own voice. There is no one-size-fits-all rule, but what matters is maintaining control and voice. If the writing no longer sounds like you, or if you are unsure whether you contributed meaningfully to the final result, it may be time to reassess your involvement in the process.

Transparency is also worth considering, especially for nonfiction or professional writers. If AI was used to structure a document or assist with phrasing, does that need to be disclosed? The answer may depend on the context. A novelist might use AI to brainstorm dialogue without informing readers, while a journalist using AI-generated summaries may be expected to disclose that fact in the interest of editorial integrity. As AI becomes more prevalent, these questions will continue to evolve, and thoughtful writers must stay engaged with the ethical implications of their choices.

Another key aspect of working effectively with AI is learning how to prompt it well. Prompts are essentially instructions, and the more specific and thoughtful they are, the better the output will be.

However, keep in mind that even strong prompting will lack a human connection.

AI output should never be accepted blindly. It should be reviewed, refined, and, where necessary, rewritten. Writers must treat AI content the way they would treat notes from a junior assistant, helpful suggestions, but not final decisions. AI may introduce clichés or reuse phrasing too often. It may overlook nuance or fail to connect ideas in a way that resonates emotionally. As the human in the process, the writer's role is to polish, adapt, and ensure that the work aligns with the intended purpose and audience.

There is also a risk of overdependence. The convenience of AI can become a crutch, reducing the writer's confidence in their own instincts and skill. Everyone knows that the more you write, the better you become. This is especially true for new writers who are still developing their voice. Relying too heavily on AI may lead to a kind of creative flattening, where the uniqueness of the creator's expression is gradually diluted by generic phrasing and predictable structure. To guard against this, writers should intentionally spend time writing without AI support. These habits keep the creative muscles strong and prevent complacency.

Working with AI also calls for humility. Sometimes the machine will suggest something you had not considered, and it will be genuinely helpful. Other times, it will offer content that feels hollow or unoriginal. Both outcomes are part of the process. Rather than reacting with fear or dependence, writers can approach AI with curiosity. *What can I learn from this? Where does this help me? Where does it fall short?* These questions turn AI from a threat into a resource, one that enhances critical thinking rather than replacing it.

In the long run, the goal is not to create like a machine, but to create alongside it, with discernment, creativity, and purpose. AI may suggest a path, but you choose where to walk. It may provide the clay, but you shape the sculpture. This partnership, when managed well, can open up new possibilities for efficiency and imagination, but it will never replace the heart of creating or leave a legacy of art that matters.

As AI becomes more advanced, the tools may change, but the responsibility remains the same. We must decide what kind of work we want to produce and how we want to produce it. How we will ensure that our voice, and not the voice of a program, remains at the center of it all. If you're smart about it, AI can make you more productive, organized, and inspired.

But it is not the creator. You are.

# CHAPTER 4

*AI and the Publishing World:*
*Disruption or Opportunity?*

The publishing world is no stranger to disruption. Over the past two decades, it has weathered the rise of self-publishing platforms, the explosion of digital books, the decline of traditional brick-and-mortar bookstores, and the increasing dominance of algorithm-driven discoverability. Currently, AI is pushing at the industry's edges. Whether viewed as an opportunity for innovation and inclusion, or a source of anxiety and fear, AI is poised to transform publishing in ways that few could have anticipated even five years ago.

But the fact is, it is.

AI in publishing, quite frankly, has been for a while. Editorial tools powered by machine learning assist with grammar checks, readability scoring, and manuscript formatting. Books are marketed through data-driven algorithms that target readers with uncanny precision. Audiobooks are being narrated by synthetic voices that are increasingly difficult to distinguish from human ones. These changes have been incremental, often slipping in quietly through tools and services that publishers and authors already use. Yet, as generative AI becomes more sophisticated, the stakes are rising. We are no longer talking solely about editing help or voice automation. We are now grappling with the reality that entire manuscripts can be written, outlined, or significantly shaped by AI systems, as well as entire audio books can be created within a few days, versus months.

This shift raises pressing questions for everyone involved in publishing. What defines authorship in a world where a book can be drafted with minimal human input? How will publishers vet authenticity, originality, and quality when manuscripts may be partially or wholly generated by machines? Can readers tell the difference, and more importantly, do they care? These questions are not hypothetical. Already, platforms have faced floods of AI-generated submissions, some of which were either plagiarized, poorly produced, or submitted in bad faith. The sheer volume of these submissions has already overwhelmed editorial teams and strained the credibility of publishing models that rely on trust and discernment.

We are gaining more and more tools, programs, and apps that can spot AI-generated material before we ever lay eyes on it, and that's important to note. We are learning key elements to look for in submissions and are daily gaining an eye for weeding out the disconnected junk.

Still, not all disruption is harmful. While AI challenges traditional definitions of creativity, it also offers tremendous potential for increasing access, expanding efficiency, and democratizing the publishing process. Independent authors can use AI tools to improve their manuscripts, design marketing strategies, or generate cover concepts without hiring large teams. Small presses with limited budgets can leverage AI to streamline formatting, generate book descriptions, or explore multiple versions of titles and taglines. These efficiencies, when used ethically and intentionally, can level the playing field for creatives who have historically lacked institutional support. Ironically, those same tools help publishers sift through authors, manuscripts, and journalists that use AI extensively.

Moving forward, the entire entertainment industry must tread carefully. The use of AI in content creation has triggered debates about copyright, intellectual property, and creative ownership. If an AI system generates a chapter based on patterns it has absorbed from thousands of existing works, does that chapter infringe on any particular author's style or phrasing? Who owns the rights to content partially generated by a machine and edited by a human? Should disclosure of AI assistance be required in published works, and if so, how should that be framed? These questions are not only legal but ethical, and they will shape the policies, contracts, and consumer expectations that govern the future of entertainment.

I know transparency will become a cornerstone issue. While most consumers may not object to light AI involvement in editing or formatting, image creating, or music production, they might feel different if they knew a novel was largely written by a machine with no human oversight. Trust relies on authenticity, even in fiction. If the emotional moments in a novel feel algorithmically generated rather than emotionally experienced, that connection may falter. The publishing industry will need to decide on this continuum and how much disclosure is owed to the audience. At the same time, avoiding knee-jerk rejection of all AI-related content, missing out on new creatives that challenge and expand existing norms.

There are also cultural implications to consider. If publishers increasingly rely on AI to determine what will sell or even to generate stories based on trend analysis, there is a risk of homogenization. The worry is that literature could become more predictable, favoring the most common structures, themes, and tropes, and reducing room for experimentation or divergence. The unique, sometimes messy, sometimes risky manuscripts that might not fit an algorithmic model could be pushed aside in favor of safer, more formulaic outputs. In this sense, overreliance on AI could lead to a narrowing of creative expression rather than an expansion of it.

Yet the outcome is not inevitable. We're learning more and more that thoughtful integration of AI into publishing practices can enhance rather than replace human judgment. We'll still use editors and agents. There are already editors that now specialize in AI detection. They are still valid and can use AI to streamline their processes. I advise authors to use tools like Grammarly or ProWriting Aid, because that doesn't negate the need for editors, but instead those tools can aid the author in cleaning up the basics before editing. An editor digs deep, reads with thought and emotion, and helps the author make their work the best that it can be on top of basic clean-ups. In addition, pre-readers are still vital, and their feedback is irreplaceable. That is because the final decisions must remain rooted in human experience and intuition. No amount of AI can replace that.

When AI was in the first stages of common use, we received a query that had promise. We asked for the rest of the fiction novel to review, and I let the author know, as we always do, that if we sign her and the manuscript goes to editing, she must be prepared to carefully consider the editors suggestions. They may want the author to scrap some sections, or flesh out other sections, etc. When the author flippantly said, "Oh yeah, that's fine. They can cut whatever they need." I knew something was amiss. Throughout my decades in publishing, I had never had that reaction. The usual responses are, "Oh no, they won't cut anything, because it's all important," "I don't want them to cut anything I've written," or "I'm leery about anything being taken out." Why do authors have that reaction? Because they've poured themselves, their blood, sweat and tears, into the manuscript. They've written, re-written, thought out every word, and revised a hundred times. It's like having a baby... and no one wants his or her baby criticized.

When one of my novels, *Love, Cutter,* came back from the first edit and the editor suggested I cut out nearly 6500 words throughout the book, I was floored. I let her know that the storyline couldn't be cut, it was important to the story. The editor replied that it might be a part of my story, but it wasn't a part of Carter's. After she said that, I read it again, through fresh eyes, and realized she was right. It was so hard to delete that storyline, because I had poured myself into it. But it was the right thing to do.

Moving onto the author that didn't care what was cut, I paused and asked, "If I send this to an editor who specializes in AI-generated material, how much of the story would it flag?" After a long pause, she admitted that much of it would be. I instructed her to go back and really dig deep. Read through it and be sure it's in her voice, relays her message, and make it more human than machine. After a few weeks of not hearing from her, I reached out, and she let me know she was too overwhelmed to do it. It required too much time she didn't have, so she was simply going to shelve the idea of publishing for now.

The bottom line is that writing a book isn't easy. As with any creative endeavor, it takes dedication, time, and thought. Readers are still moved by stories that surprise them, challenge them, and reflect truths that cannot be captured by code alone. The publishing industry's role, as always, is to steward these stories, to bring them into the world with care, and to protect the space where human imagination thrives, not fully AI-generated content that falls flat.

I want to say to you. If you are writing a book just to have a book, then you are in it for the wrong reason, anyway. What do you want to say? What inspired you to want to put this story into the world? Are you hoping to help someone who went through what you did? Or write out an entertaining novel because it keeps bouncing around in your head? Do you want to help someone overcome their fear of using technology? Analyze your own process and why you want it.

Writers and all creators alike must play a part in shaping this future. By using AI ethically and transparently, they model a responsible relationship between creativity and technology. By maintaining ownership over their voice and setting clear boundaries where AI assists rather than authors, writers help sustain trust with their readers. Video and image creators maintain trust with their viewers. Musicians and producers maintain trust with their listeners. By pushing back against the temptation to use AI as a shortcut to mediocrity, we all can uphold the value of thoughtful, original, and resonant storytelling.

In the end, AI is neither a savior nor a destroyer of producing and publishing. Just like with any technology, we will adapt. AI should be viewed as a tool, one with immense power and equally immense responsibility. Its impact will be determined not by its capabilities alone, but by the choices we all make in response to it. Disruption, after all, is not new to any industry, especially the entertainment industry. What matters is how an industry responds, whether it retreats in fear or adapts with intention.

The opportunity before the publishing world is real. AI can expand access, improve workflows, and empower new voices. But it must be balanced with clarity, with creativity, and with a relentless commitment to human stories that have always made books worth reading. The ways may evolve, the methods may shift, and the tools may change, but the core purpose remains... stories still matter. No matter how intelligent a machine becomes, it will never replace the soul of the person wielding it.

# CHAPTER 5

## *The Hidden Cost of Smart Tools*

Every new tool we adopt has a footprint. We don't always see it. It doesn't always make noise. But it's there. Artificial intelligence, for all its wonder and usefulness, is no different. I am not simply talking about the digital footprint, which, for AI-generated writing, music composition, vocals, image, or animation, all have their own hidden footprints. I am talking about our physical footprint.

I've talked a lot about AI in terms

of creativity and how it can support our process without stealing our voice. But there's another side of the conversation that deserves attention. Not of the philosophical kind, or even of the personal-moral kind. This is a practical, real-world issue that's easy to overlook in all the excitement of possibility: the environmental cost.

AI, especially in the way it's expanding now, uses a staggering amount of energy. When you prompt a chatbot or run an image generator, you don't feel the weight of that. You don't see the data centers that make it happen. But they're there. Huge facilities, often the size of football fields, powering our questions, our edits, and our output.

Most people don't realize that data centers don't just use energy, they use an *incredible amount of water*. Water is essential for cooling the massive machines that run AI. On hot days especially, when electricity demand is already high, these centers pull in even more water to prevent overheating. In some regions, this can mean millions of gallons each year, water that's diverted from already-stressed ecosystems and communities. While we're busy creating content, some of these systems are quietly straining one of the planet's most precious resources.

It's not unlike the early days of the microwave, which my family embraced with full curiosity and a little bit of awe. I remember us sitting in a class together, learning how to use it properly. It was cutting-edge back then. But it's one thing to learn how to use a microwave. It's another to understand what it means to build millions of them and plug them into a growing, power-hungry world. Yet, most of us use them.

AI runs on the backs of advanced chips, often made from rare earth metals mined at great environmental cost. Then there's the electricity. Training large AI models isn't like booting up your laptop. It's more like powering an entire neighborhood. Some estimates compare a single training run of a major AI model to the carbon footprint of flying multiple people across the world. That's just the training. Every time we ask AI to generate something, that too takes energy. Multiply that by the millions of prompts being run every day, and the numbers grow fast.

Of course, this isn't the first time a new

technology has brought new concerns. My dad used to say, "With every jump in tech, we lose something and gain something. The trick is knowing which is which." He was never afraid of what was new, but he never stopped asking questions either.

That mindset has stuck with me. I'm not sharing this chapter to spread fear or shame anyone into unplugging their computer, by far. I'm sharing it because I believe we should all walk into innovation with our eyes wide open.

The good news is that this conversation is already underway. Many tech companies are investing in more sustainable energy sources for their data centers, like solar, wind, and hydroelectric. Others are developing ways to train models more efficiently, using less power and fewer resources. Researchers are working on smaller models that can do the same work without the same environmental load. Some AI companies have committed to becoming carbon neutral or even carbon negative.

But here's where we come in. Every time we use a tool like AI, we have a chance to pause and ask ourselves: Is this a place where I really need it? Or is this a moment I could create on my own, unplugged?

This isn't about guilt. It's about mindfulness. Responsibility. Just like fast fashion or fast food, fast content can come at a cost if we're not careful. Like with any industry, we have to be willing to adapt, to use these tools responsibly, to advocate for transparency, to support companies that are doing the work to reduce their footprint.

AI isn't going anywhere. It will only get smarter, faster, and more embedded in our daily lives. But as creators, as storytellers, thinkers, teachers, and business owners, we get to choose how we use it and how we care for the world that hosts it.

So, let's stay curious. Let's stay honest. Let's stay human, even in a world more and more powered by machines.

Because technology doesn't get to decide our values. We do.

# CHAPTER 6

## *The Dark Side of Easy Content*

The promise of artificial intelligence in the world of writing is both dazzling and dangerous. On the one hand, AI tools can accelerate the creating process, eliminate technical errors, and help us brainstorm out of creative ruts. On the other hand, as I've brushed on earlier, they can seduce creators into a relationship with content that favors quantity over quality, speed over depth, convenience over intention, and, much more importantly, *trend and current mindset over core beliefs and values.* This seduction, while subtle, poses a growing threat to the integrity of creative work. The dark side of easy content is not only the erosion of originality but also the increasing temptation to outsource thinking, emotion, and even authorship to machines.

Creators have always grappled with pressure. There are deadlines to meet, platforms to populate, audiences to satisfy, and algorithms to please. The modern digital landscape rewards frequency and visibility. In this climate, AI tools offer an alluring temptation. A writer no longer needs to wrestle with every sentence, because the machine can generate thousands of words in minutes. It can produce headlines, social media captions, blog posts, outlines, summaries, and promotional emails without breaking a sweat. For businesses and creators operating under tight timelines or low budgets, this efficiency is transformative.

However, when the goal becomes to produce more and more content with less and less effort, the work begins to lose something essential: humanity.

What is lost is often subtle at first. It

may begin with a blog post that sounds slightly more generic than usual or a piece of dialogue that lacks the spark of lived emotion. Over time, if the writer continues to rely on AI without infusing the work with human insight, the voice begins to flatten. Originality fades. The stories start to feel hollow, derivative, or uninspired. The writing is grammatically sound, but it doesn't have any spark. The words do not surprise, challenge, or resonate. They merely fill space.

There is a growing risk of plagiarism, intentional or not. Generative AI tools rely on vast amounts of preexisting data. They analyze patterns in language drawn from books, articles, websites, and other sources, many of which are copyrighted or proprietary. While the output may be technically new, it is often a patchwork of styles and ideas that echo existing works. This raises serious ethical questions. When a creator uses AI to generate content, can they be sure the result is truly original? If a passage closely mirrors the work of another author, who bears responsibility? The user, the machine, or the developer?

I've heard it said many times that

no thought is original to man, and yes that is true. But delivery is everything. How many people thought of inventing something, but never took the steps to do so, and then are shocked when someone else does?

It happens all the time.

These questions are not hypothetical. Several well-known platforms have already faced backlash for publishing AI-generated material that appeared to borrow too heavily from human-written texts. In some cases, entire articles or stories have been flagged for similarity to previously published works. This not only undermines trust in the content but also threatens the reputation of those who use the tools carelessly. All creators who want to maintain credibility must approach AI with awareness, acknowledging that convenience should never override responsibility.

I want to be clear, you aren't replacing someone else's job when you use AI to fully generate your content, you are replacing *yourself*. That is eye-opening when you consider the implications. Would you let your mother replace your voice? Your friend? Current culture? No, you want your own voice, and we want your voice as well. If generating content was what entertainment was all about, we wouldn't need authors at all. We could simply hire anyone, give them a theme or summary of a work we want to publish, and pay them to generate it. However, that is not what we, or the world, wants or needs. We need human connection, relatability, and heartfelt entertainment.

As more creators and companies turn to AI to generate writing quickly and cheaply, digital platforms are becoming flooded with repetitive, thoughtless material. Search engines, social media feeds, and online marketplaces are increasingly populated by content that looks polished but offers little substance.

One of the main worries is that the glut of content can bury meaningful work, making it harder for readers to find writing that is nuanced or emotionally rich. When everyone is publishing constantly, attention becomes the most precious commodity. When attention is scarce, depth is often sacrificed.

A writer's voice is a fragile, hard-earned thing. It develops over time through reflection, experimentation, and failure. It is shaped by memory, emotion, and lived experience. When writers begin to rely too heavily on AI for phrasing, structure, or tone, their voice begins to blur. Their writing starts to sound like everything else, smooth but indistinct. Polished but poor. Common and unoffensive. Readers may not immediately notice what is missing, but they will feel it. They will sense the words aren't not quite alive. In a world awash with content, it is the human voice that cuts through the noise. Losing it is not just a creative risk, but a personal one.

I am not saying that AI must be avoided entirely. On the contrary. I hate outlining. I don't even usually do it with fiction. Very often I just start writing. However, with nonfiction, I feel it's a must. To be able to plug my chapter titles or previous blog posts titles I am using in the book into an AI platform and get an outline within seconds, is the part of this technology that I enjoy. On one of my books, the generator told me that one chapter would flow better within the content if it were moved up two chapters. Something my editors and I didn't catch, and all agreed with wholeheartedly. So, my goal is absolutely not to tell you to reject technology, but to use it with discernment. Remain conscious of what you are trading when you choose convenience. You must ask whether the use of AI serves the work or simply expedites it. You must reflect on what you want your writing to be and why you began writing in the first place. Is it to connect, to explore, to challenge, or simply to fill a page?

Just as with any technological advancement, the tools themselves are not the enemy. The danger lies in how easily those tools can be misused.

# CHAPTER 7

*The Ethics of Authorship*

One of the most pressing and complex issues is the question of authorship in the evolving world of AI. As more writers, entrepreneurs, educators, marketers, and even casual content creators incorporate AI into their work, the traditional understanding of what it means to be an "author" is being challenged. Authorship has always implied more than mere ownership of words. It suggests originality, intentionality, and accountability. It indicates that the writer has not only produced the work but has taken responsibility for its voice, its truth, and its impact. In the AI landscape, the ethical dimension of authorship becomes harder to define and even harder to uphold.

To begin with, it is essential to ask what it means to author something in the age of AI. If a writer feeds a prompt into an AI system and receives a fully formed essay or chapter in return, who is the true author? The person who wrote the prompt, or the system that composed the result? Some might argue that the user deserves credit, having initiated the process, and currently that is true if you pay for a service. Others might see this as insufficient involvement, especially if the majority of the work was generated without significant human contribution, editing, or revising. The dilemma intensifies when that content is presented to readers who assume, often without question, that a human mind and heart shape every sentence.

We're currently asking the questions. When AI is used to fully generate content, to what extent should the viewer be informed? In academic writing, journalism, or nonfiction, the use of AI may soon be a required disclosure, particularly when factual accuracy, attribution, or credibility is at stake.

In fiction, poetry and songwriting, where creativity and emotional resonance take precedence, the expectation of full human authorship has always been implied. If a novel or short story has been significantly written by AI tools, should that be stated as well, even though it is fiction?

While no universal standards currently exist, these questions are becoming increasingly urgent, and we soon expect there to be a standard to follow throughout the entire entertainment industry.

While I am focusing the majority of this book on writers, I also want to address other aspects of AI throughout. When it comes to AI-generated images, the question is often, "Did the person sending the prompt to generate the image have to edit?" If so, was it minimal or extensive?

In the US, copyright protection typically doesn't extend to content created solely by AI, as human authorship is a requirement. In January 2025, the U.S. Copyright Office released a report (Part 2) clarifying its position on AI-generated works and copyright, stating that current copyright law can apply. Judicial precedent also confirms that human authorship is a requirement.

Copyright protection isn't lost simply by using AI as a creative aid. The presence of a human author's distinct creative expression and control in the final product is a factor in its copyrightability.

In this report it states, "The Office affirms that existing principles of copyright law are flexible enough to apply to this new technology, as they have applied to technological innovations in the past. It concludes that the outputs of generative AI can be protected by copyright only where a human author has determined sufficient expressive elements. This can include situations where a human-authored work is perceptible in an AI output, or a human makes creative arrangements or modifications of the output, but not the mere provision of prompts."

Shira Perlmutter, Register of Copyrights and Director of the U.S. Copyright Office furthers states within the report, "After considering the extensive public comments and the current state of technological development, our conclusions turn on the centrality of human creativity to copyright, Where that creativity is expressed through the use of AI systems, it continues to enjoy protection."

The preliminary Part 3 of the study states in conclusion, "Throughout its history, copyright law has adapted to new technology, furthering its progress while preserving incentives for creative activity. This has enabled our nation's creative and technology industries to become global leaders in their fields. While the use of copyrighted works to power current generative AI systems may be unprecedented in scope and scale, the existing legal framework can address it as in prior technological revolutions. The fair use doctrine in particular has served to flexibly accommodate such change. We believe it can do so here as well."

To understand this simply, if you have a book that is 100% AI generated, even through your own prompts, it is not copyrightable.

Again, technology changes, and we should adapt accordingly.

These tools are exciting and are growing

by leaps and bounds. They offer platforms for creators and artists who have never had opportunities before. But they still must have a human behind them. To go back to my dad's belief, we can all be excited about many of the advances, but also considerate of them.

Beyond legal definitions, there is a moral responsibility that accompanies authorship. Writers and creators alike are cultural contributors. Their works influence others, shape conversations, and reflect shared human experiences. This responsibility extends to intent. Why is AI being used, and what does the creator or author hope to gain from it? If the goal is to enhance clarity, improve grammar, or organize complex information, the use of AI may align with the broader commitment to communication and quality.

If, however, the goal is to shortcut the creative process, inflate content volume, or mimic emotional resonance without investing emotional effort, then the ethical line begins to blur.

I have several unpublished children's books, as well as a pile of poetry and songs that stands two feet or more stacked on top of one another, some dating back to

1980, when I was ten years old. Much of that creativity and work had sat dormant for years. They were personal, like my books were before publishing. A mark of my ever-racing brain. In 2018, my novel, *Two Thousand Lines*, had just come back from editing, and in it the main character talks about a book her mom used to read to her. One day, after reading through the manuscript before publishing, my husband noticed that the main characters were all based on doodles he'd done through all the years. I loved them and kept them in our picture chest for many years. So, it was a nod to him, of course. He asked me if I was planning to do the children's book also, a question he fully regrets ever asking, because I decided then, as a companion book to the novel, I would create the children's book and publish them both the same day. Although reluctant, he agreed to let me use his doodles. After the first picture book was complete, I thought of several other topics that could help kids face common issues such as jealousy, fitting in, and isolation and The *Hollows on the Bayou* series was born. Even though the graphics are simple, these books took months to create. I knew that no matter how many children's books I'd already written in the past, they'd stay tucked away, because I simply didn't have the time or money to create any more of them.

Therefore, my other children's books sat dormant. In publishing, I've helped many people get their kids' books to print through the company but knew doing my own would require time I didn't have. Until AI. Even though I buy character art and commercial licenses from artists for ethical reasons, and the AI-generated images still need a ton of work (because, well, they're AI), it's still way faster now to create the images I need for the books. Artificial intelligence provided the space to create something I never would've been able to focus on for myself and opened a space for authors who never would've been considered through the company.

That's where AI helps authors, artists, and publishers. Using it ethically, responsibly, and still putting in the work makes assistance from artificial intelligence worth noting.

We have to be honest with ourselves about whether we are using AI to serve our work or to avoid the work entirely.

This self-awareness is vital in maintaining integrity and trust with the public.

Another ethical dimension lies in the question of labor and equity. As AI-generated content becomes more common, there is a growing concern that human writers and producers, particularly those working as freelancers or content creators, may be devalued. That companies and clients seeking quick and inexpensive content might opt for AI-generated material rather than paying a human to create it. However, currently, a human still has to be behind the action. Then the question is how much human is behind it? This shift can reduce opportunities for creative professionals, lower industry standards, and create a race to the bottom in terms of quality and compensation. Ethical authorship must therefore consider not only individual integrity but collective impact. Creators who use AI should be mindful of how their choices influence the broader creative economy and what precedent they may be setting for the future of work.

Going back earlier with the children's books, understand, this did not take away jobs from illustrators, because I never would've completed those projects. Instead, many character illustrators are getting their work purchased that weren't before. Their work is reaching a broader base, and they are making more money. In addition, the vocation is expanding because many book illustrators are now specializing in AI-generated content based on legally purchased characters and elements, which opens a new space for their artistic contribution.

The entertainment industry as a whole must change and adapt as it has for every major technologic advancement.

To move forward ethically, we must establish personal and professional guidelines. These might include setting limits on how much AI can be involved in a project, committing to disclosing AI assistance when appropriate, and continuously refining our understanding of what constitutes meaningful authorship. In time, industry standards will emerge, offering clearer frameworks for responsible use. Until then, the burden lies with each of us to determine how to balance innovation with integrity.

Ultimately, authorship in the age of AI is not about rejecting technology but about reclaiming intention. It is about choosing to remain fully present in the creative process, even when machines offer to take over. It is about valuing not just the final product but the journey of thought, emotion, and reflection that gives writing its meaning. I believe creators do not need to fear AI, but they do need to remain vigilant in their voice. Remember that while a machine can assemble words and images, it cannot offer insight or vision. It cannot take responsibility. And it cannot speak with the authority of someone who has lived, who has felt, who has dared to shape silence into a story. It is to say, this is mine, not because I claim ownership over every word, beat, or image, but because I stand behind what this work represents of me. AI can assist, inspire, and even collaborate, but it cannot replace the act of standing.

In the end, ethics in authorship is not about rules but about relationships between the creator, their work, and their readers, listeners, or consumers.

# CHAPTER 8

## *The Shame and Early Judgement of Innovation*

There's a strange pattern that repeats every time a new tool enters the creative world. First comes skepticism. Then comes fear. Then, for many who dare to adopt it, comes shame.

A subtle, unspoken guilt. A fear that using it makes the work "less real," or makes them "less of a creator." As someone who has watched technology evolve across every part of the creative industry, I want to call that out and name it for what it is: just another chapter in the long story of how creativity and innovation collide and expand.

We've been here before.

As one example, we saw this happen when digital audio workstations (DAWs) like FL Studio, originally called "Fruity Loops," hit the music scene. Traditional producers scoffed at it. It was said that real musicians didn't "click beats," they played instruments. The software was mocked, often dismissed as amateur or "garage studio junk." Fruity Loops was seen as a toy. An amateur's shortcut. Something you messed around with in your bedroom or garage, not something you built a career on. Yet, that so-called toy launched careers.

Massive ones.

Producers like Metro Boomin, Avicii, Martin Garrix, Skrillex, and countless others crafted entire hits from those early digital audio workstations. They didn't need a full studio. They needed access, curiosity, and a tool that didn't judge them for not having a $20,000 setup. DAWs opened a door that had been closed for a long time. Like every meaningful shift in technology, it didn't ruin the craft, it changed who had access to it and how to improve it. The software, and others like it, made music production accessible to thousands of young creatives who didn't have access to expensive studios or classical training. It also enhanced the industry. What started as "not real music" helped redefine music altogether.

Before the DAWs, the controversy surrounded other innovations such as multitrack recording and the synthesizer, all of which are now used regularly and widely acknowledged as industry norms. Each of these was thought to cause the downfall of the music industry and believed that all music would become "canned" or irrelevant. Which is the opposite. Music exploded, expanded, and evolved.

We've come to believe easier and authentic are opposites. Let me just add, FL Studios is not "easy." I have used it for years, since before the rebranding, and it is challenging. However, for me, it was easier than learning an instrument, or two or three, because I don't have the aptitude for that, but I do have the aptitude for technology. And that's okay. We are all good at something. Understanding, learning, and using DAWs still requires plenty of time and effort. That's my lane. Also, now, with the assistance of AI technology, I'm now better equipped to bring my lyrics and vision to life.

This fear is not new. We've done this same dance every time a new tool emerges. When the typewriter replaced the pen, writers were accused of losing intimacy with their words. When photography emerged, painters declared it the death of art. When drum machines hit the mainstream, traditional drummers worried they'd become obsolete. When Photoshop entered the design world, people said it made real artists irrelevant. When microwave ovens hit the market, people worried no one would go to restaurants anymore. *Spoiler: none of these predictions came true.*

The common denominator in all of these fears was a perceived loss of control, money, or norms. A belief that ease equals erasure. That if something gets easier, it must also get worse. But ease is not the enemy. Misuse is.

The fear behind the shame is often tied to job security or human negation. If a machine can produce something "good enough," why would anyone pay for the "real" thing? Why would anyone hire a composer, a copywriter, a video editor, a poet? Why would anyone go to a restaurant when they could have a meal in minutes at home? Will this be the end of us as humans? Will we become irrelevant? But history has shown again and again that new tools do not eliminate the need for humans. They expand the ways in which we create, live, and enjoy our lives, as well as expand our marketplace.

The internet didn't eliminate authors; it gave rise to self-publishing. Auto-Tune didn't destroy vocals; it gave birth to a whole new soundscape. CGI didn't end cinema; it gave us entire worlds we couldn't previously imagine. Microwaves didn't illuminate restaurants; they gave them tools to offer additional better, quicker options, reduce food waste, as well as offer families more time together around the table. These innovations all created entirely new career opportunities and products.

In each of these transitions, early adopters faced scrutiny, mockery, and even public shaming. But in time, the conversation shifted. In many instances, what was once judged as inauthentic became the new industry standards. Not because the tool was perfect, but because the people using it were committed to using it well.

The same will be true of AI.

However, public shame of innovation feeds off the fear that you'll be found out, that someone will say, "You didn't do this the hard way." But the harder or "normal" way is not always the right way. Technology growth has proven this time and again, because it has always existed to reduce friction, not to replace soul and hard work. Each person must make an individual choice to use or not to use.

There is no shame in using a GPS to get to a destination. There is no shame in using spellcheck to catch your typos. There is no shame in using AI to outline your ideas or generate alternatives when you're stuck. Shame only enters when we confuse the tool with the work and the public scrutiny with the reality. Freedom comes when we replace the fear with the progress.

The heart of any creation still has to come from you.

Let me offer this: Needing support

does not make you less of a creator. Does a parent microwaving dinner make them bad? No, however, many people believed it did way back when they first hit the consumer. Thankfully, I had parents who didn't place a lot of value in what other people thought was best for them. They made decisions on what was best for their family and didn't fear technology.

Ironically, the people who ridicule others for using AI? They've more than likely used it themselves. Or something just like it. Maybe it was spellcheck. Maybe it was a smart plugin. Maybe it was Google itself. We are all already using augmented tools, even if we pretend otherwise.

In time, this current discomfort with AI will look a lot like the early days of DAWs, or the skepticism toward digital publishing, or the distrust of online influencers. It will fade. Because as more great work is produced through these tools, as more readers are moved, more books, movies, and music are made, and more lives are positively impacted, the shame will fade away.

Yes, we must remain ethical. Yes, we must guard originality and protect our voices and others. But we must also recognize that the tools we use do not define us, the way we use them does.

It begs to mention that some AI music generators have trained their models on vast amounts of information, some copyrighted. So, be careful when using any generator. Be sure you edit to fit your vision and not just accept what is created through prompts alone. AI training means it is trained on previous works, so unfortunately some sites may use actual parts of existing works, and you do not want that. Do not prompt author and artists names. Be ethical. Shame thrives in secrecy if you aren't being ethical. Keep in mind also that there are affordable DAW systems out there for your use.

If you are being ethical and have been hiding the fact that AI helped you sketch out your outline, create vocal for your lyrics, summarize your ideas, or polish your work, stop hiding. You are creating. You're thinking, shaping, building.

You're human.

If we continue to let shame police innovation, we will silence voices that were finally finding their way into the room.

*Let's not do that.*

Instead, let's remember what history has shown us again and again: new tools always come with controversy, but the ones who use them responsibly will always rise above it.

Don't apologize for growing with the tools that are helping you create. Don't let shame stop you from doing the thing you were meant to do.

While I do believe in regulation, we must remember there is no honor in creative gatekeeping. There is power in creative courage as the tools keep evolving.

Let's evolve with them, honestly, responsibly, ethically, and without shame.

# CHAPTER 9

*Augmented Intelligence, Not Artificial*

As the conversation around artificial intelligence continues to expand across industries, creative fields are learning to distinguish between what is artificial and what is augmented. The term "artificial intelligence" often conjures images of machines replacing people, robots taking over creative work, or human value being pushed aside by faster, more efficient algorithms. Yet, a more accurate and hopeful framing for writers and artists is "augmented intelligence." This term shifts the focus from replacement to enhancement, from competition to collaboration. When approached through the lens of augmentation, AI becomes not a rival to creativity but a powerful ally, one that can enhance us do what we already do.

The distinction is not merely semantic; it is philosophical. Artificial intelligence suggests something separate from or in opposition to human thought. It implies an independent entity that mimics or replaces the functions of the mind. Augmented intelligence, on the other hand, positions AI as an extension of human capability, a tool designed to support our thinking and doing, not to replacing our thinking or doing. Writers, artists, and creators who view AI in this way are more likely to use it purposefully, selectively, and ethically. They see the technology not as a threat but as a resource, one that works best when paired with human insight, emotion, and critical reasoning.

There is also a practical side to augmentation that cannot be ignored. For many creators, especially those managing multiple responsibilities, AI can help preserve time and mental energy. A self-published author might use AI to help draft marketing blurbs, write metadata, or outline a press release. A teacher preparing lesson plans or a freelancer juggling multiple clients might use AI to streamline administrative writing. A songwriter might use AI to help with music composition. When AI takes on some of this workload, creators are able to devote more attention to the creative or personal projects that matter most to them, giving some a new avenue they might never have otherwise.

Even so, the use of AI as an enhancing force depends heavily on the creator's awareness and control. Without a clear sense of personal boundaries, what begins as extension can quickly turn into substitution and duplication. We must remain vigilant in maintaining our presence within the work. It is only beneficial when it supports human creativity, not when it replaces the creative struggle, the reflection, or the decision-making that defines our authorship.

In education and mentorship, the concept of augmented intelligence has particular value. Emerging writers can use AI tools to better understand the mechanics of language, test out different narrative structures, or compare stylistic approaches. Instructors can guide students in using AI to learn and develop, rather than to shortcut their development.

Professionally, augmented intelligence can also help level the playing field for creators with learning differences, physical disabilities, or language barriers. Dictation tools powered by AI can assist those who struggle with traditional typing. Grammar and clarity checkers can support those writing in a second language. Concept organizers and summarization tools can help those with ADHD or executive function challenges stay on track. In this context, AI is not a threat to creativity but a support system that empowers more people to express themselves and share their stories.

Still, there are boundaries worth protecting. We must be careful not to confuse convenience with conscience.

The future of creating and writing involves artificial intelligence, whether we like it or not. How that future unfolds depends on how we choose to engage with it. If we treat AI as a replacement for thought, a shortcut to success, or a way to avoid creative labor, we risk diminishing the art form. If instead we treat AI as a tool for discovery, refinement, empowerment, and opportunity, we can elevate the work we do and deepen the impact we make.

When technology becomes part of that process, it must serve that mission, not distract from it. It should help us bring our works to life, not replace the life within them.

# CHAPTER 10

## *Finding Your Balance: The Human-AI Partnership*

There are no legal guidelines yet that structure how it can be used and reproduced aside from laws that already exist, although I imagine that will come soon. AI generators are already placing markers or identifiers within the output for future recognition as laws and guidelines evolve. There is no universal formula yet for this balance, no fixed ratio of how much AI involvement is acceptable and how much must come directly from the writer's mind, but it is coming.

Currently, balance begins with **clarity**. Setting personal rules to help us stay grounded. Some may choose to never allow AI to write dialogue, since capturing the subtleties of human interaction requires intuition, context, and emotional nuance. Others might decide that plot structure can be loosely influenced by AI suggestions, as long as the final decisions reflect their vision. These guidelines are not meant to limit creativity but to protect it. Boundaries foster intentionality, and intentionality is what keeps AI in its proper place as a tool, not a creator.

Another important aspect of balance is **pacing**. The speed with which AI can generate content is both beneficial and hazardous. As an example, writers accustomed to spending hours crafting a single paragraph may find the instant production of multiple drafts liberating. However, this speed can also encourage a rush to publish, bypassing the necessary steps of reflection, revision, and refinement. Writing that endures often takes time. It requires consideration, layering, and moments of pause. AI can assist with many things, but it cannot replace the slow, iterative process through which deep insight and strong voice emerge. Writers who prioritize speed over substance may find themselves with more content but far less connection.

We must also regularly **evaluate** how AI use affects our confidence and skill development. While AI can serve as a helpful guide, excessive reliance may erode a creator's capabilities. If every sentence is rewritten by a machine, if every paragraph is structured by a template, the writer begins to lose their actual voice, which is often the reason they started writing in the first place. This is particularly true for newer writers who are still discovering their style and learning to trust their voice. To find balance, writers who find they are often relying on AI, must set it aside and challenge themselves to create without it.

It is equally important to **listen** for changes in voice. One of the most common risks of frequent AI use is the gradual flattening of tone. Because AI draws from patterns found in existing text, its output often defaults to a neutral, broadly appealing style that lacks distinct personality. If left unchecked, this influence can begin to seep into the writer's own language, softening the edges of their voice, rounding out the sharpness of their observations, and smoothing over the idiosyncrasies that made their work compelling. To counteract this effect, writers should regularly reread their older work and notice whether their voice is evolving or fading. They should ask whether their newest writing still sounds like them, still carries the rhythms, quirks, and sensibilities that define their identity on the page.

Balance also means being willing to **reject** AI suggestions. Just because a machine offers a smoother transition or a more marketable headline does not mean it is the right choice. We must remember our message and our values. If an AI-generated sentence does not align with those elements, it should be rewritten personally or discarded. There is no shame in refusing a tool's advice. In fact, that refusal is a sign of strength, a signal that we are still in command. AI is not infallible. It does not understand context the way humans do. It cannot measure the weight of a word or the timing of a phrase in the same way a human reader can.

Therefore, its suggestions must always be filtered through human discernment.

To further support balance, we can **build** rituals around our use of AI. For instance, a writer might start a writing session with an hour of free writing before opening any AI tools, ensuring that their own thoughts take precedence. They might use AI in timed blocks, returning to manual work for reflection or deeper composition. These rituals help anchor the writer in their own mind and reduce the risk of defaulting to automation.

Keep in mind that writing or creating works is not only about production; it is about *presence.*

Community can also play a role in maintaining balance. Creators who **share** their process with others, who talk openly about how they use AI and what they struggle with, are more likely to develop thoughtful practices. Learning from each other's successes and mistakes, borrowing strategies, and setting collective norms. As the creative landscape continues to change, community becomes our source of accountability and encouragement. In the current state, creators can feel shame about using AI, so they often don't want to disclose the use of it. But we have to grow with the times, and keep an open space for it.

Finally, balance requires ongoing **reflection**. Technology will continue to evolve, and we will need to adjust our practices in response. What feels appropriate today may not feel right tomorrow. We must periodically revisit our goals, asking whether our use of AI still aligns with the kind of work we want to create.

# CHAPTER 11

## *A Future Worth Writing*

In the midst of rapid technological change, writers and creators stand at a crossroads. Artificial intelligence is no longer a future concept or speculative tool; it is part of the daily landscape. Creators are using it to brainstorm, to draft, to revise, and to promote. Businesses are adopting it to streamline communication, educators are turning to it for lesson planning and personalized learning, and publishers are experimenting with it to manage workloads and meet deadlines. Music engineers, while having used forms of it in the past, are using it more and more to generate compositions, or enhance sound. The question is no longer *whether AI will become part of the creative process* but how deeply it will be allowed to shape the work and the people behind it. With so much possibility at hand, the most pressing challenge for us all is not technical adaptation, but creative responsibility.

Authenticity is no longer a vague aspiration or artistic flourish; *it is a defining feature of relevance.* People crave connection, not just information. In the post pandemic world of disconnect, that is truer than ever.

Those that view AI as a competitor may take solace in the fact that true voices, true connection, and a higher standard of creating are here. AI is a part of our current and future, not as a competitor, but as a collaborator, because it also serves as a filter. While publishers, editors, and readers are more and more able to spot overuse of AI by simply reading, we now also, again ironically due to AI, have programs that can root out many works that are mostly AI generated.

About a year ago, one of the manuscript excerpts that was submitted for publishing was compelling. So, we asked for more and sent that off to one of our storyline editors who immediately sent it back, noting that much of the work was AI generated. When I asked the writer about this, they were initially shocked that we knew it, then extremely embarrassed. They admitted that they did use AI programs after they got writer's block. They admitted after using it to flesh out a paragraph, how easy it was to simply plug in their thoughts and quickly generate the rest of the chapters they'd summarized. I explained that while the first half of the manuscript was compelling, much of the rest fell absolutely flat. I advised them to go back, reread, and rewrite it in their own voice. It took them nearly six months, but when they resubmitted it, the manuscript was engaging, emotional, and cathartic.

So, to hammer my point home, when treated as a tool, not a replacement, AI can help reduce barriers to entry, expand access to creative expression, and support creators in producing more intentional work.

Creators who understand how to use AI well will not lose their identity or voice; they will find new ways to sharpen it. The machine may suggest possibilities, but the writer will still choose which story to tell, which sentence or musical note to keep, and which truth to pursue, and revisit to reflect their message.

The temptation to hand over responsibility to the machine will grow stronger as tools become more powerful and convenient. It will be easy to justify shortcuts, to delegate one task after another, and to convince oneself that speed and efficiency are signs of progress. But remember, writing has never been merely about productivity. It is about reflection, imagination, and discovery. It is a conversation between self and world, between thought and language. When writing becomes too easy, it often loses that sense of dialogue. It becomes less of an offering and more of a transaction. A future worth writing is one in which creators still value the struggle, still embrace the process, and still honor the intimacy of putting something personal into the work.

Since AI-generated content has become

more common, the risk of deception has increased. Writers who choose to use AI ethically can choose to communicate that choice clearly, either through disclaimers, acknowledgments, or behind-the-scenes transparency with publishers.

We must all continue to refine our craft. The availability of AI should not become an excuse to create less thoughtfully. On the contrary, it should motivate us to deepen our skills, to explore language more fully, and to bring greater intentionality to our work. This opportunity should not be wasted. The future depends on people who are willing to grow as artists, not just as content creators.

Publishing and creative industries must also rise to the occasion. If companies continue to prioritize speed, volume, and profit over quality, authenticity, and innovation, the creative landscape will suffer. Consumers will become desensitized to formulaic content as creators will feel pressure to conform to machine-generated norms. Instead, the industry should support voices that take time, that reflect nuance, that challenge assumptions.

I always try to make the time, even

if it's inconvenient and not business savvy, for writers to write under their own timeline, not ours. When a writer says they have writer's block, I always correct them, because for years I have said writer's block is simply writer's think. If you never take the time to think through what you are writing, you are wasting precious time. Thinking through an idea, storyline, or even a chapter, is one of the most important parts of the writing process. If you never thought of the book, you'd never be writing it. That same principle follows you through the entire writing process, and should.

A dear friend of ours is a best-selling author with a list of best-selling books. One of his main processes is getting alone to think. Don't devalue that part and feel stuck in it. Embrace it!

While AI has provided a more inclusive environment, I see that in the very near future as AI takes more common place, regardless of AI fears, that writing will become more exclusive and more courageous. It will welcome technology but not surrender to it. Creators, the ones who still dedicate the time, thought, and energy into their vision, will embrace new possibilities while honoring timeless values of the work. True voices will endure and be highlighted. Overused or 100% AI-generated material will be evident and largely rejected. Because it is human intention that gives art its shape, its strength, and its soul.

Let's build a future in which technology helps us devote more time to *devote more fully, more freely, and more faithfully* to who we are.

# CONCLUSION

## *The Person Still Matters*

At every stage of this journey, one truth remains clear: AI is not the enemy of creativity unless we allow it to be. The rise of AI in writing, publishing, image and video generation, movie and song production, brings with it a wave of opportunity, challenge, and reflection. Creators now have access to tools that can accelerate ideation, improve their work, and support the technical aspects.

This book is not a manual for machine-made content, nor is it a warning of creative apocalypse. It is an invitation to find balance. It is a reminder that technology, no matter how advanced, is still just a tool, and the hand that holds it matters. AI can support the structure, but it cannot offer the soul. It can suggest a sentence, but it cannot live the story behind it.

That work remains ours.

In this new age, the future belongs to those who choose intentionality over automation, authenticity over efficiency, and expression over imitation. The ones who understand that using AI to enhance their work does not mean surrendering their identity in the process.

I believe we do not have to fear the tools of advancing technology, but to protect your voice, insight, vision, and commitment. Whether you write books, blogs, business plans, or journal entries, or create images, videos, music composition, or enhance movie production, the question is not simply what you can produce, but what you are trying to say, and why it matters that you are the one saying it.

Let the machines offer their speed and structure. Let the prompts provide their push. But let your heart continue to shape the story.

The world may be changing, but the true, soulful, engaging heart of a person who bleeds their thoughts and creativity into their works, still matters.

And always will.

# LETTER TO YOU

## *From the Author*

As a publisher and creator, I have the privilege and the responsibility of working with a wide range of authors, artists. editors, agents, and industry professionals.

Over the past several years, I've watched with growing curiosity and concern as artificial intelligence has dramatically entered the creative space in a very real way. Some are using these tools thoughtfully and ethically, allowing AI to support their work without overshadowing it. Others, however, are leaning too far into convenience, relying on AI to do the heavy lifting they once did with intention, care, and heart.

This began as a small handbook I created for some of our authors and artists, a companion guide to help them navigate this new landscape with confidence and clarity. I handed it out at meetings, tucked it into welcome packets, and referenced it in workshops. Eventually, I realized I wanted to share to a broader audience.

This is a practical, honest conversation about how people can use AI without losing themselves, their work, others' works, and our glorious world of creativity in the process. It is an invitation to stay grounded, creative, and above all, to stay human.

Whether you are a writer, an educator, a content creator, a production engineer, or a lifelong learner, I hope this book serves you well. I hope it helps you use technology as a tool, not a crutch. My deepest hope is it reminds you that your voice, your experience, and your intention still matter more than anything a machine can generate.

If you're asking: *Did she use AI to assist her with this book?* The answer is, yes, I absolutely did. On purpose, responsibly, and ethically, using my own voice, my own thoughts, and my own previous work on the subject to help me create something that can help you do the same. The base cover image was AI-generated through prompts, and many images later, and a trip to graphic design for plenty of editing, the cover was born.

Happy creating.

*With respect for the craft and belief in your voice,*

*Michelle Jester*

Author and Publisher

# ABOUT THE AUTHOR

Michelle Jester is a former Crisis Counselor and Public Relations Consultant, in conjunction with working in publishing for twenty-plus years. She spends her time writing in addition to maintaining her full-time publishing career. An author of fiction, nonfiction, and children's books, Michelle is also a contributing author to the #1 bestseller, My Labor Pains Were Worse than Yours.

She lives in Louisiana with her husband, high school sweetheart and retired Master Sergeant. Michelle wears a bracelet every day with a single, yellow, rubber duckie charm on it to remind her to enjoy the fun and happy things of life!

Feel free to connect with her, and sign up for her monthly newsletter: MichelleJester.net

# ACKNOWLEDGEMENTS

I am so thankful to the many people who support me, most of all my husband, who is by far my biggest fan, inspiration, idea developer, and critic. Without him, I wouldn't have the freedom to be authentically me and go wherever my mind takes me.

To my best friends, Wendy, Yvette, and Tracie, thank you for always encouraging me and my work. Having you three supportive women around me is one of my most valued treasures.

To my children, thank you both for being my creative sounding boards, and listening to my ideas for hours on end. I am proud of you both for being you and following God.

Thanks to all the editors, graphic designers, pre-readers, as well as all the app and program creators for the many avenues you provide creators to enhance, produce, and finalize our work.

# REFERENCES

### Historical Tech Disruption (Betamax case)

- Sony Corp. v. Universal City Studios ("Betamax case"), decided January 17, 1984. Supreme Court ruled that noncommercial home recording was fair use and VCRs had legitimate purposes

  Data Matters Privacy BlogFilmBuffOnline+9Justia Law+9Encyclopedia.com+9.

  https://supreme.justia.com/cases/federal/us/464/417/

- *Oyez summary,* "Sony Corporation of America v. Universal City Studios, Inc." (accessed via Oyez).

  https://www.oyez.org/cases/1982/81-1687 Justia Law+1Legal Information Institute+1Justia Law+3Oyez+3Wikipedia+3

- *Justia overview,* same case details: Decided Jan 17 1984, fair-use ruling upheld

  Westport Tech Museum+3Legal Information Institute+3Wikipedia+3.

### AI & Copyright—U.S. Copyright Office Report (Jan 29, 2025)

- U.S. Copyright Office, *Copyright and Artificial Intelligence,*

Part 2: *Copyrightability*, issued January 29, 2025. Clarifies that AI-generated works can receive copyright only with substantial human creative input

ASCAP+7U.S. Copyright Office+7The Library of Congress+7.

https://www.copyright.gov/newsnet/2025/1060.html

• Copyright Office Releases Part 2 of *Artificial Intelligence Report*

https://www.copyright.gov/newsnet/2025/1060.html

Issue No. 1060 - January 29, 2025

• Copyright Office Releases Part 3 of *Artificial Intelligence Report*

https://www.copyright.gov/policy/artificial-intelligence/

Preliminary Pending Publication

## Environmental Cost of AI & Data Centers

• *IEA* (International Energy Agency), AI driving steep data center electricity demand—set to double by 2030 to ~945 TWh annually

U.S. Copyright Office+2U.S. Copyright Office+2Copyright Lately+2Wikipedia+14IEA+14Wikipedia+14.

https://www.iea.org/news/ai-is-set-to-drive-surging-electricity-demand-from-data-centres-while-offering-the-potential-to-transform-how-the-energy-sector-works

• MIT News, *Explained: Generative AI's environmental impact,* January 17, 2025—data centers reached ~460 TWh in 2022, could approach 1,050 TWh by 2026; North American AI servers doubled power usage from 2022 to 2023 IEA+1MIT Sloan+1MIT News.

https://news.mit.edu/2025/explained-generative-ai-environmental-

- Planet Detroit (citing 2021 study): U.S. data centers use ~7,100 L water per MWh; Google data centers consumed ~12.7 billion liters in 2021 MIT NewsPlanet Detroit.

  https://planetdetroit.org/2024/10/ai-energy-carbon-emissions/

- Wikipedia, "Data center"—a 100 MW facility may consume ~2,000,000 L of water per day; global footprint ~560 billion L/year doubling by 2030

  Planet DetroitWikipedia.

  https://en.wikipedia.org/wiki/Data_center

- Wikipedia, "Environmental impact of artificial intelligence" (2025), estimating AI water withdrawals of 4.2 6.6 billion $m^3$ by 2027; training GPT 3 consumed ~700,000 L

  reuters.com+6Wikipedia+6Business Insider+6Wikipedia+1arXiv+1.

  https://en.wikipedia.org/wiki/Environmental_impact_of_artificial_intelligence

### AI Energy Demand & Sustainability Solutions

- Time.com, *The AI Revolution Isn't Possible Without an Energy Revolution*, June 2025—Sam Altman testifies that AI growth limited by energy, U.S. needs ~90 GW additional capacity

  WikipediaTIME.

  https://time.com/7294803/ai-revolution-energy-revolution/

- TechRadar, *How can we create a sustainable AI future?* (July 2025)—AI/data centers at 1–2 % global electricity demand (~500 TWh by 2030); highlights liquid cooling, renewable siting

  TIMEtechradar.com.

  https://www.techradar.com/pro/how-can-we-create-a-sustainable-ai-future

- Business Insider, June 2025 story on Digital Realty upgrading AI-capable data centers with renewable & liquid cooling, reducing water and emissions

techradar.com+1techradar.com+1Business Insider.

https://www.businessinsider.com/digital-realty-ai-infrastructure-data-centers-sustainability-strategy-2025-6

- IEEE PSU, blog on AI's growing energy/water footprint: U.S. data center electricity ~4.4 % in 2023, could triple by 2028

Business Insideriee.psu.edu.

https://iee.psu.edu/news/blog/why-ai-uses-so-much-energy-and-what-we-can-do-about-it

### Academic Studies: Water & Carbon Footprint

- ArXiv (2023), *Making AI Less "Thirsty"*: GPT 3 training consumed ~700,000 L water; global AI water use projected 4.2–6.6 billion m³ by 2027

iee.psu.eduarXiv.

https://arxiv.org/abs/2304.03271

- ArXiv (2025), *How Hungry is AI?*: GPT 4o query uses ~0.43 Wh; 700 million queries consume as much electricity as 35,000 homes and evaporate water for 1.2 M people

arXivarXiv.

https://arxiv.org/abs/2505.09598

Having trouble writing that book? Feel you don't have the time? Don't know what's stopping you? Where to start? It can sometimes seem like that great book idea keeps getting further and further away. Learn how to overcome sticking points and get that book done.

Many people dream of writing a book, but often struggle to make it a reality. With our busy lives, common obstacles like not knowing where to start or how to find the time arise repeatedly. Even if you've started writing, you may feel as though writer's block or self-doubt is stopping you.

With twenty-plus years of experience in publishing, Michelle Jester, a former Crisis Counselor and Public Relations Consultant, dedicates herself to helping authors realize their dream. In addition to her full-time career in publishing, she commits herself to writing using her name and different pseudonyms. An author of fiction, nonfiction, and children's books, Michelle is also a contributing author to several bestsellers.

Having helped many people through the process of writing, publishing, and marketing, Michelle is excited to share her How-to books with you. Now, you can finally make that great idea a reality and start writing that book you've always dreamed of.

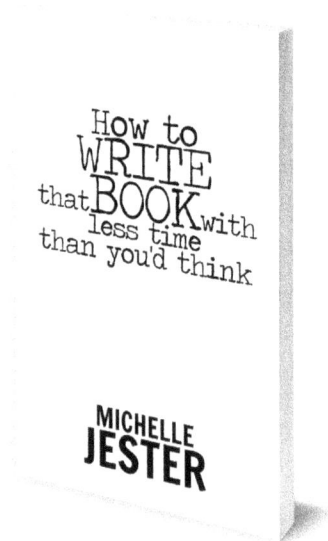

How to
WRITE
that BOOK with
less time
than you'd think

MICHELLE
JESTER

www.ingramcontent.com/pod-product-compliance
Lightning Source LLC
Chambersburg PA
CBHW060623210326
41520CB00010B/1446